# The Orchard Files

Preserve Your Orchard Knowledge, Cultivation Practices, and Memories for Future Generations

AF880917

## Property of:

________________________________________

________________________________________

ISBN: 978-1-968971-24-3

Then God said, “I give you every seed-bearing plant on the face of the whole earth and every tree that has fruit with seed in it. They will be yours for food.”

Genesis 1:29

Ironstone Tennessee

Disclaimer and Terms of Use:

The author and publisher have made every effort to ensure the completeness of this record book. However, they do not guarantee that all information is error-free or entirely comprehensive.

While diligent efforts have been made to verify the content, the author and publisher are not responsible for any errors, omissions, or differing interpretations of the material presented. Any unintended references that may appear to slight individuals, businesses, or organizations are purely coincidental.



Copies of this book are available for book signings, special events, and retail gift shop purchases. For direct ship information, of copies of 10 or more, please contact the author at ironstone_tennessee@yahoo.com.

The Orchard Files
Preserve Your Orchard Knowledge, Cultivation Practices, and Memories for Future Generations

**ISBN: 978-1-968971-24-3**
Printed in the United States of America.

First printing, 2025

Ironstone Tennessee

Calhoun Tennessee, 37309

www.ironstone-tennessee.com

# Notes

 

# Table of Contents

# How to Use The Orchard Files

**I want to thank you so much for choosing to purchase The Orchard Files. I welcome feedback and suggestions for ways to make it more useful for future editions.**

You can add stickers, doodle in the margins, add tabs, and make notes as needed. When you're finished, you'll have created a valuable reference as you go forward. This reference can serve as a treasured heirloom to pass on to future generations of gardeners, or a priceless resource to pass on to the future owners of your property.

## Important Dates and Data:

Record your **USDA growing zone, average and observed frost dates, average annual rainfall, soil type and pH here.** This information allows you to choose plants that will be most likely to do well in your climate. Most plant labels and catalog listings will tell you the plant's range of growing zones.

Knowing your **average first and last frost dates** will help you to schedule appropriate shipping dates when purchasing live plants and trees online. It will also help you to know when it is (probably) safe to transplant outdoors.

Knowing your annual average precipitation (snow and rain) will help you to choose what plants to grow, based on your desired amount of time spent in the garden (and how big of a water bill you can afford). If you are a lazy gardener like me, or just don't have time in for your schedule for frequent watering, then choosing trees that can thrive with only rainfall (most of the time) makes sense.

Your **soil type and pH** will determine what amendments you may need, and what plants will and won't do well for you in the garden. A bit of quick Internet research will tell you whether a plant likes your type of soil and whether your soil pH is in the right range for it to thrive. Knowing this can save you a lot of money and work.

 

## *My Gardening Challenges*:

These are factors that you can't change easily (or at all), such as a short growing season, your work schedule, soil type, physical limitations, small budget or climate. While it's easy to think of these challenges as a big negative, I would encourage you to instead think of them as boundaries that help you to keep your focus. When you know from the start that there are things you can't do, you are free to focus instead on the things you can do, and then make a plan to do those things well. Which brings us to:

## *My Gardening Goals*:

Are you crazy for fresh peaches? Do you live for apple butter on fresh, homemade biscuits? Do you want to grow a year's supply of strawberries and blueberries for your breakfast smoothies? Having goals and keeping them in focus will help you to achieve the orchard of your dreams without going broke in the process.

## *Harvest and Pruning Charts*:

The purpose of the harvest and pruning charts is to give you a space to see when your workload is the heaviest throughout the course of the year. They are organized into 3 months per page. The Harvest Month pages allow you to further see what each month's harvest schedule looks like, week-by-week.

## *Sections*:

There are 9 sections designed for specific types of fruits and nuts. Apples-Pears-Quince, Stone Fruits, Nuts, Citrus, Berries, Vines, Native Plants, Tropicals, and Caffeine-Producing Shrubs. Each section provides pages designed for specific types for fruit or nut trees and shrubs. Most people won't use all the sections, but the Files are designed to be comprehensive. If you find that you do not have enough pages for a specific type of plant, please email me at: ironstone-tennessee@yahoo.com and provide a screen shot of your purchase receipt and the name of the plant that you need more pages for. I'd be happy to send you a PDF so that you can print more copies of the page you need.

 

## Orchard Files:

The blank line under “”Apple Files”, “Peach Files” etc. is where you can write in the name of the cultivar such as “Gala” or “Belle of Georgia”.

## Size Type or Thorns:

Size refers to the root stock for grafted fruit and nut trees. They may be dwarfing, semi-dwarf or standard height. Type can refer to things like high bush, low bush or rabbit eye blueberries, or early/mid/late-season or day-neutral strawberries. Many cultivars of citrus and brambles have thorns, but some do not.

## Fruit or Leaf Qualities:

Some apples are great for cider, while others are best for pies, sauce, fresh eating or winter storage. Some peaches are best for canning while others are best eaten fresh. China tea leaves are smaller and less cold hardy than Assam tea leaves.

## Bloom Time:

You can be specific or general. A tree's bloom time in relation to your last frost date will influence how likely you are to harvest fruit in any given year. If a late frost hits when the blossoms are vulnerable, your fruit tree may not produce any fruit that year. Many growers will choose both early and late blooming varieties to increase their chances of a harvest.

## Cross-Pollinator:

Many fruit trees require another cultivar to ensure good pollination, so it’s important to choose cultivars that bloom at the same time.

## Harvest Time:

Depending on your needs, you might prefer to plant your strawberries for a long harvest season, or you may want a couple of weeks of intense harvesting and then done for the season. Some apples are ripe in summer, and others aren't ready until late October. Knowing when to expect your harvest helps with scheduling and prevents wasted fruit.

### *Source*:

It's good to record where your trees and shrubs come from for several reasons. If you decide you want more, you know where to go. If you aren't happy with a plant's performance, you'll know to try a different source next time. If someone asks you where you got those delicious blackberries, you can, if you choose, share your source.

### *Disease Resistance*:

Many universities are working to develop fruit trees that are resistant to common diseases and pests. If your area is prone to cedar apple rust, for example, you might want to make sure you plant apple trees that are resistant to that fungal disease.

### *Pruning*:

For most fruit trees and shrubs, there is an ideal time to prune, and it's not the same time for everything. If you don't know when that it, look it up and then make note of it here. And then, make sure that note gets added to your gardening calendar.

### *Soil Amendments*:

Some plants like blueberries require the application of fertilizers or pH adjustment from time to time. Record the plant's needs here so you can find the information quickly, and again, make sure that note gets on your gardening calendar!.

### *Diseases and Treatments*:

Some cultivars of plants are more disease prone than others. For those fruits and nuts are not disease-resistant, you can usually find a treatment or prevention method that fit your gardening philosophy. Make note of that here.

### *Pest and Treatments*:

Most plants have some kind of insect or animal that loves to feast on them. If you know what to watch out for, you can make a plan to prevent, treat or protect your plants as you see fit. You can also decide whether not it's worth your time to grow those plants that are more vulnerable to pests.

 

**Notes:**

The notes area is for your observations, ideas, plans, recollections and thoughts you don't want to forget. Things like: "Grandma made the best pies from these peaches, need to find her recipe" or "Japanese beetles hardly touched this apple tree, maybe plant more."

## I wish you Fruitful Gardening!

# Important Dates and Data

| Growing Zone | Soil Type | Soil pH | Avg. Last Frost | Avg. 1st Frost | Avg. Rainfall |
|---|---|---|---|---|---|
| | | | | | |

My Gardening Challenges:

My Gardening Goals:

 

# Harvest Times

| Month/Week | January | | | | February | | | | March | | | |
|---|---|---|---|---|---|---|---|---|---|---|---|---|
| Plant | 1 | 2 | 3 | 4 | 1 | 2 | 3 | 4 | 1 | 2 | 3 | 4 |
| | | | | | | | | | | | | |
| | | | | | | | | | | | | |
| | | | | | | | | | | | | |
| | | | | | | | | | | | | |
| | | | | | | | | | | | | |
| | | | | | | | | | | | | |
| | | | | | | | | | | | | |
| | | | | | | | | | | | | |
| | | | | | | | | | | | | |
| | | | | | | | | | | | | |
| | | | | | | | | | | | | |
| | | | | | | | | | | | | |
| | | | | | | | | | | | | |
| | | | | | | | | | | | | |
| | | | | | | | | | | | | |
| | | | | | | | | | | | | |
| | | | | | | | | | | | | |
| | | | | | | | | | | | | |
| | | | | | | | | | | | | |
| | | | | | | | | | | | | |
| | | | | | | | | | | | | |
| | | | | | | | | | | | | |
| | | | | | | | | | | | | |
| | | | | | | | | | | | | |
| | | | | | | | | | | | | |
| | | | | | | | | | | | | |

 

# Harvest Times

| | Month/Week | April | | | | May | | | | June | | | |
|---|---|---|---|---|---|---|---|---|---|---|---|---|---|
| Plant | | 1 | 2 | 3 | 4 | 1 | 2 | 3 | 4 | 1 | 2 | 3 | 4 |
| | | | | | | | | | | | | | |
| | | | | | | | | | | | | | |
| | | | | | | | | | | | | | |
| | | | | | | | | | | | | | |
| | | | | | | | | | | | | | |
| | | | | | | | | | | | | | |
| | | | | | | | | | | | | | |
| | | | | | | | | | | | | | |
| | | | | | | | | | | | | | |
| | | | | | | | | | | | | | |
| | | | | | | | | | | | | | |
| | | | | | | | | | | | | | |
| | | | | | | | | | | | | | |
| | | | | | | | | | | | | | |
| | | | | | | | | | | | | | |
| | | | | | | | | | | | | | |
| | | | | | | | | | | | | | |
| | | | | | | | | | | | | | |
| | | | | | | | | | | | | | |
| | | | | | | | | | | | | | |
| | | | | | | | | | | | | | |
| | | | | | | | | | | | | | |
| | | | | | | | | | | | | | |
| | | | | | | | | | | | | | |
| | | | | | | | | | | | | | |
| | | | | | | | | | | | | | |

 

# Harvest Times

| Month/Week | July | | | | August | | | | September | | | |
|---|---|---|---|---|---|---|---|---|---|---|---|---|
| **Plant** | **1** | **2** | **3** | **4** | **1** | **2** | **3** | **4** | **1** | **2** | **3** | **4** |
| | | | | | | | | | | | | |
| | | | | | | | | | | | | |
| | | | | | | | | | | | | |
| | | | | | | | | | | | | |
| | | | | | | | | | | | | |
| | | | | | | | | | | | | |
| | | | | | | | | | | | | |
| | | | | | | | | | | | | |
| | | | | | | | | | | | | |
| | | | | | | | | | | | | |
| | | | | | | | | | | | | |
| | | | | | | | | | | | | |
| | | | | | | | | | | | | |
| | | | | | | | | | | | | |
| | | | | | | | | | | | | |
| | | | | | | | | | | | | |
| | | | | | | | | | | | | |
| | | | | | | | | | | | | |
| | | | | | | | | | | | | |
| | | | | | | | | | | | | |
| | | | | | | | | | | | | |
| | | | | | | | | | | | | |
| | | | | | | | | | | | | |
| | | | | | | | | | | | | |
| | | | | | | | | | | | | |
| | | | | | | | | | | | | |

 

# Harvest Times

| Month/Week | October | | | | November | | | | December | | | |
|---|---|---|---|---|---|---|---|---|---|---|---|---|
| **Plant** | 1 | 2 | 3 | 4 | 1 | 2 | 3 | 4 | 1 | 2 | 3 | 4 |
| | | | | | | | | | | | | |
| | | | | | | | | | | | | |
| | | | | | | | | | | | | |
| | | | | | | | | | | | | |
| | | | | | | | | | | | | |
| | | | | | | | | | | | | |
| | | | | | | | | | | | | |
| | | | | | | | | | | | | |
| | | | | | | | | | | | | |
| | | | | | | | | | | | | |
| | | | | | | | | | | | | |
| | | | | | | | | | | | | |
| | | | | | | | | | | | | |
| | | | | | | | | | | | | |
| | | | | | | | | | | | | |
| | | | | | | | | | | | | |
| | | | | | | | | | | | | |
| | | | | | | | | | | | | |
| | | | | | | | | | | | | |
| | | | | | | | | | | | | |
| | | | | | | | | | | | | |
| | | | | | | | | | | | | |
| | | | | | | | | | | | | |
| | | | | | | | | | | | | |
| | | | | | | | | | | | | |
| | | | | | | | | | | | | |

 

# January Harvest

| Week 1 |
| --- |
| |
| |
| |
| |
| |
| |
| |
| |
| |
| |

| Week 2 |
| --- |
| |
| |
| |
| |
| |
| |
| |
| |
| |
| |

| Week 3 |
| --- |
| |
| |
| |
| |
| |
| |
| |
| |
| |
| |

| Week 4 |
| --- |
| |
| |
| |
| |
| |
| |
| |
| |
| |
| |

 

# February Harvest

| Week 1 |
| --- |
| |
| |
| |
| |
| |
| |
| |
| |
| |
| |

| Week 2 |
| --- |
| |
| |
| |
| |
| |
| |
| |
| |
| |
| |

| Week 3 |
| --- |
| |
| |
| |
| |
| |
| |
| |
| |
| |
| |

| Week 4 |
| --- |
| |
| |
| |
| |
| |
| |
| |
| |
| |
| |

 

# March Harvest

## Week 1

## Week 2

## Week 3

## Week 4

 

# April Harvest

| Week 1 |
| --- |
| |
| |
| |
| |
| |
| |
| |
| |
| |
| |

| Week 2 |
| --- |
| |
| |
| |
| |
| |
| |
| |
| |
| |
| |

| Week 3 |
| --- |
| |
| |
| |
| |
| |
| |
| |
| |
| |
| |

| Week 4 |
| --- |
| |
| |
| |
| |
| |
| |
| |
| |
| |
| |

 

# May Harvest

| Week 1 |
| --- |
| |
| |
| |
| |
| |
| |
| |
| |
| |
| |

| Week 2 |
| --- |
| |
| |
| |
| |
| |
| |
| |
| |
| |
| |

| Week 3 |
| --- |
| |
| |
| |
| |
| |
| |
| |
| |
| |
| |

| Week 4 |
| --- |
| |
| |
| |
| |
| |
| |
| |
| |
| |
| |

 

# June Harvest

## Week 1

## Week 2

## Week 3

## Week 4

 

# July Harvest

## Week 1

## Week 2

## Week 3

## Week 4

 

# August Harvest

## Week 1

## Week 2

## Week 3

## Week 4

 

# September Harvest

## Week 1

## Week 2

## Week 3

## Week 4

 

# October Harvest

## Week 1

## Week 2

## Week 3

## Week 4

 

# November Harvest

## Week 1

## Week 2

## Week 3

## Week 4

 

# December Harvest

| Week 1 |
|---|
| |
| |
| |
| |
| |
| |
| |
| |
| |
| |

| Week 2 |
|---|
| |
| |
| |
| |
| |
| |
| |
| |
| |
| |

| Week 3 |
|---|
| |
| |
| |
| |
| |
| |
| |
| |
| |
| |

| Week 4 |
|---|
| |
| |
| |
| |
| |
| |
| |
| |
| |
| |

 

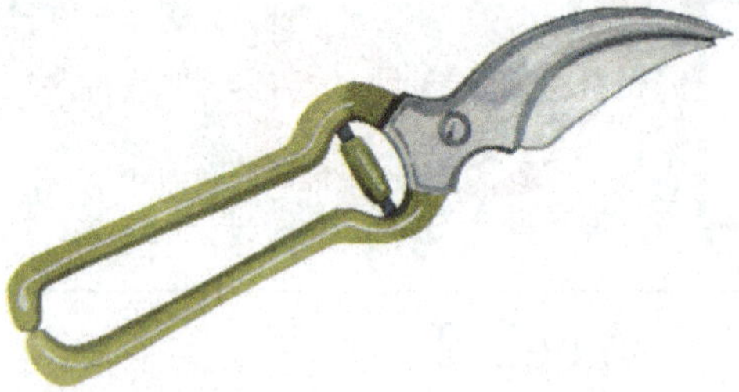

# *Pruning Times*

| Month/Week | January | | | | February | | | | March | | | |
|---|---|---|---|---|---|---|---|---|---|---|---|---|
| **Plant** | **1** | **2** | **3** | **4** | **1** | **2** | **3** | **4** | **1** | **2** | **3** | **4** |
| | | | | | | | | | | | | |
| | | | | | | | | | | | | |
| | | | | | | | | | | | | |
| | | | | | | | | | | | | |
| | | | | | | | | | | | | |
| | | | | | | | | | | | | |
| | | | | | | | | | | | | |
| | | | | | | | | | | | | |
| | | | | | | | | | | | | |
| | | | | | | | | | | | | |
| | | | | | | | | | | | | |
| | | | | | | | | | | | | |
| | | | | | | | | | | | | |
| | | | | | | | | | | | | |
| | | | | | | | | | | | | |
| | | | | | | | | | | | | |
| | | | | | | | | | | | | |
| | | | | | | | | | | | | |
| | | | | | | | | | | | | |
| | | | | | | | | | | | | |
| | | | | | | | | | | | | |
| | | | | | | | | | | | | |
| | | | | | | | | | | | | |
| | | | | | | | | | | | | |
| | | | | | | | | | | | | |
| | | | | | | | | | | | | |

 

# Pruning Times

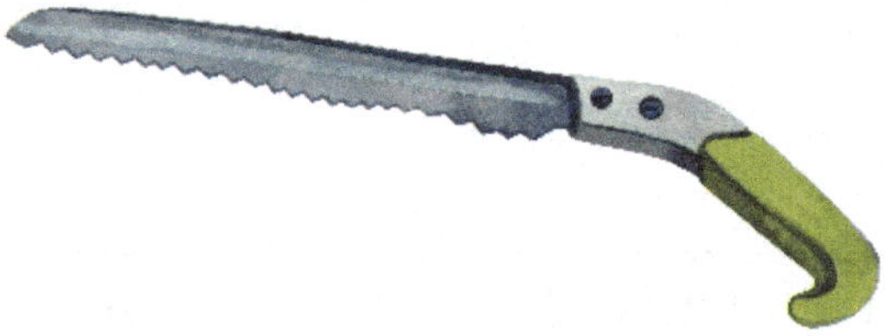

| Month/Week | April | | | | May | | | | June | | | |
|---|---|---|---|---|---|---|---|---|---|---|---|---|
| **Plant** | **1** | **2** | **3** | **4** | **1** | **2** | **3** | **4** | **1** | **2** | **3** | **4** |
| | | | | | | | | | | | | |
| | | | | | | | | | | | | |
| | | | | | | | | | | | | |
| | | | | | | | | | | | | |
| | | | | | | | | | | | | |
| | | | | | | | | | | | | |
| | | | | | | | | | | | | |
| | | | | | | | | | | | | |
| | | | | | | | | | | | | |
| | | | | | | | | | | | | |
| | | | | | | | | | | | | |
| | | | | | | | | | | | | |
| | | | | | | | | | | | | |
| | | | | | | | | | | | | |
| | | | | | | | | | | | | |
| | | | | | | | | | | | | |
| | | | | | | | | | | | | |
| | | | | | | | | | | | | |
| | | | | | | | | | | | | |
| | | | | | | | | | | | | |
| | | | | | | | | | | | | |
| | | | | | | | | | | | | |
| | | | | | | | | | | | | |
| | | | | | | | | | | | | |
| | | | | | | | | | | | | |
| | | | | | | | | | | | | |

 

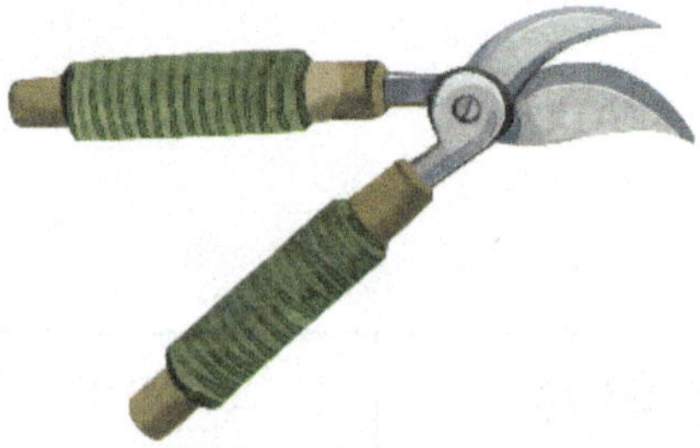

# Pruning Times

| Month/Week | July | | | | August | | | | September | | | |
|---|---|---|---|---|---|---|---|---|---|---|---|---|
| **Plant** | **1** | **2** | **3** | **4** | **1** | **2** | **3** | **4** | **1** | **2** | **3** | **4** |
| | | | | | | | | | | | | |
| | | | | | | | | | | | | |
| | | | | | | | | | | | | |
| | | | | | | | | | | | | |
| | | | | | | | | | | | | |
| | | | | | | | | | | | | |
| | | | | | | | | | | | | |
| | | | | | | | | | | | | |
| | | | | | | | | | | | | |
| | | | | | | | | | | | | |
| | | | | | | | | | | | | |
| | | | | | | | | | | | | |
| | | | | | | | | | | | | |
| | | | | | | | | | | | | |
| | | | | | | | | | | | | |
| | | | | | | | | | | | | |
| | | | | | | | | | | | | |
| | | | | | | | | | | | | |
| | | | | | | | | | | | | |
| | | | | | | | | | | | | |
| | | | | | | | | | | | | |
| | | | | | | | | | | | | |
| | | | | | | | | | | | | |
| | | | | | | | | | | | | |
| | | | | | | | | | | | | |
| | | | | | | | | | | | | |

 

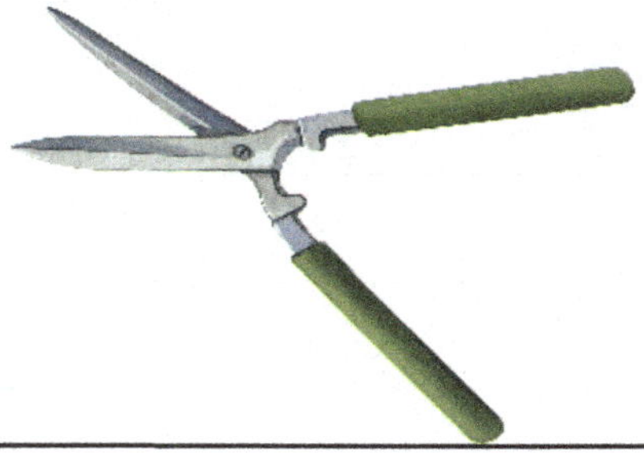

# Pruning Times

| Month/Week | October | | | | November | | | | December | | | |
|---|---|---|---|---|---|---|---|---|---|---|---|---|
| **Plant** | **1** | **2** | **3** | **4** | **1** | **2** | **3** | **4** | **1** | **2** | **3** | **4** |
| | | | | | | | | | | | | |
| | | | | | | | | | | | | |
| | | | | | | | | | | | | |
| | | | | | | | | | | | | |
| | | | | | | | | | | | | |
| | | | | | | | | | | | | |
| | | | | | | | | | | | | |
| | | | | | | | | | | | | |
| | | | | | | | | | | | | |
| | | | | | | | | | | | | |
| | | | | | | | | | | | | |
| | | | | | | | | | | | | |
| | | | | | | | | | | | | |
| | | | | | | | | | | | | |
| | | | | | | | | | | | | |
| | | | | | | | | | | | | |
| | | | | | | | | | | | | |
| | | | | | | | | | | | | |
| | | | | | | | | | | | | |
| | | | | | | | | | | | | |
| | | | | | | | | | | | | |
| | | | | | | | | | | | | |
| | | | | | | | | | | | | |
| | | | | | | | | | | | | |
| | | | | | | | | | | | | |
| | | | | | | | | | | | | |

 

# Notes

 

# Apples, Pears & Quince

# Apple Files

| | |
|---|---|
| Size | |
| Fruit Qualities | |
| Bloom Time | |
| Cross-Pollinator | |
| Harvest Time | |
| Source | |
| Disease Resistance | |
| Pruning | |
| Soil Amendments | |
| Diseases & Treatments | |
| Pests & Treatment | |

**Notes:**

# Apple Files

______________________________

| | |
|---|---|
| Size | |
| Fruit Qualities | |
| Bloom Time | |
| Cross-Pollinator | |
| Harvest Time | |
| Source | |
| Disease Resistance | |
| Pruning | |
| Soil Amendments | |
| Diseases & Treatments | |
| Pests & Treatment | |

**Notes:**

# Apple Files

| | |
|---|---|
| Size | |
| Fruit Qualities | |
| Bloom Time | |
| Cross-Pollinator | |
| Harvest Time | |
| Source | |
| Disease Resistance | |
| Pruning | |
| Soil Amendments | |
| Diseases & Treatments | |
| Pests & Treatment | |

**Notes:**

# Apple Files

| | |
|---|---|
| Size | |
| Fruit Qualities | |
| Bloom Time | |
| Cross-Pollinator | |
| Harvest Time | |
| Source | |
| Disease Resistance | |
| Pruning | |
| Soil Amendments | |
| Diseases & Treatments | |
| Pests & Treatment | |

**Notes:**

# Apple Files

_______________

| | |
|---|---|
| Size | |
| Fruit Qualities | |
| Bloom Time | |
| Cross-Pollinator | |
| Harvest Time | |
| Source | |
| Disease Resistance | |
| Pruning | |
| Soil Amendments | |
| Diseases & Treatments | |
| Pests & Treatment | |

**Notes:**

# *Apple Files*

______________________

| | |
|---|---|
| Size | |
| Fruit Qualities | |
| Bloom Time | |
| Cross-Pollinator | |
| Harvest Time | |
| Source | |
| Disease Resistance | |
| Pruning | |
| Soil Amendments | |
| Diseases & Treatments | |
| Pests & Treatment | |

**Notes:**

# Apple Files

| | |
|---|---|
| Size | |
| Fruit Qualities | |
| Bloom Time | |
| Cross-Pollinator | |
| Harvest Time | |
| Source | |
| Disease Resistance | |
| Pruning | |
| Soil Amendments | |
| Diseases & Treatments | |
| Pests & Treatment | |

**Notes:**

# Apple Files

| | |
|---|---|
| Size | |
| Fruit Qualities | |
| Bloom Time | |
| Cross-Pollinator | |
| Harvest Time | |
| Source | |
| Disease Resistance | |
| Pruning | |
| Soil Amendments | |
| Diseases & Treatments | |
| Pests & Treatment | |

**Notes:**

# Apple Files

| | |
|---|---|
| Size | |
| Fruit Qualities | |
| Bloom Time | |
| Cross-Pollinator | |
| Harvest Time | |
| Source | |
| Disease Resistance | |
| Pruning | |
| Soil Amendments | |
| Diseases & Treatments | |
| Pests & Treatment | |

**Notes:**

# Apple Files

| | |
|---|---|
| Size | |
| Fruit Qualities | |
| Bloom Time | |
| Cross-Pollinator | |
| Harvest Time | |
| Source | |
| Disease Resistance | |
| Pruning | |
| Soil Amendments | |
| Diseases & Treatments | |
| Pests & Treatment | |

**Notes:**

# Apple Files

| | |
|---|---|
| Size | |
| Fruit Qualities | |
| Bloom Time | |
| Cross-Pollinator | |
| Harvest Time | |
| Source | |
| Disease Resistance | |
| Pruning | |
| Soil Amendments | |
| Diseases & Treatments | |
| Pests & Treatment | |

**Notes:**

# Apple Files

| | |
|---|---|
| Size | |
| Fruit Qualities | |
| Bloom Time | |
| Cross-Pollinator | |
| Harvest Time | |
| Source | |
| Disease Resistance | |
| Pruning | |
| Soil Amendments | |
| Diseases & Treatments | |
| Pests & Treatment | |

**Notes:**

# Pear Files

| | |
|---|---|
| Size | |
| Fruit Qualities | |
| Bloom Time | |
| Cross-Pollinator | |
| Harvest Time | |
| Source | |
| Disease Resistance | |
| Pruning | |
| Soil Amendments | |
| Diseases & Treatments | |
| Pests & Treatment | |

**Notes:**

# Pear Files

| | |
|---|---|
| Size | |
| Fruit Qualities | |
| Bloom Time | |
| Cross-Pollinator | |
| Harvest Time | |
| Source | |
| Disease Resistance | |
| Pruning | |
| Soil Amendments | |
| Diseases & Treatments | |
| Pests & Treatment | |

**Notes:**

# Pear Files

| | |
|---|---|
| Size | |
| Fruit Qualities | |
| Bloom Time | |
| Cross-Pollinator | |
| Harvest Time | |
| Source | |
| Disease Resistance | |
| Pruning | |
| Soil Amendments | |
| Diseases & Treatments | |
| Pests & Treatment | |

**Notes:**

# Pear Files

| | |
|---|---|
| Size | |
| Fruit Qualities | |
| Bloom Time | |
| Cross-Pollinator | |
| Harvest Time | |
| Source | |
| Disease Resistance | |
| Pruning | |
| Soil Amendments | |
| Diseases & Treatments | |
| Pests & Treatment | |

**Notes:**

# Quince Files

_______________________________

| | |
|---|---|
| Size | |
| Fruit Qualities | |
| Bloom Time | |
| Cross-Pollinator | |
| Harvest Time | |
| Source | |
| Disease Resistance | |
| Pruning | |
| Soil Amendments | |
| Diseases & Treatments | |
| Pests & Treatment | |

**Notes:**

# Quince Files

| | |
|---|---|
| Size | |
| Fruit Qualities | |
| Bloom Time | |
| Cross-Pollinator | |
| Harvest Time | |
| Source | |
| Disease Resistance | |
| Pruning | |
| Soil Amendments | |
| Diseases & Treatments | |
| Pests & Treatment | |

**Notes:**

 

# Notes

 

# Stone Fruits

# Apricot Files

| | |
|---|---|
| Size | |
| Fruit Qualities | |
| Bloom Time | |
| Cross-Pollinator | |
| Harvest Time | |
| Source | |
| Disease Resistance | |
| Pruning | |
| Soil Amendments | |
| Diseases & Treatments | |
| Pests & Treatment | |

**Notes:**

# Apricot Files

| | |
|---|---|
| Size | |
| Fruit Qualities | |
| Bloom Time | |
| Cross-Pollinator | |
| Harvest Time | |
| Source | |
| Disease Resistance | |
| Pruning | |
| Soil Amendments | |
| Diseases & Treatments | |
| Pests & Treatment | |

**Notes:**

# Apricot Files

| | |
|---|---|
| Size | |
| Fruit Qualities | |
| Bloom Time | |
| Cross-Pollinator | |
| Harvest Time | |
| Source | |
| Disease Resistance | |
| Pruning | |
| Soil Amendments | |
| Diseases & Treatments | |
| Pests & Treatment | |

**Notes:**

# Apricot Files

| | |
|---|---|
| Size | |
| Fruit Qualities | |
| Bloom Time | |
| Cross-Pollinator | |
| Harvest Time | |
| Source | |
| Disease Resistance | |
| Pruning | |
| Soil Amendments | |
| Diseases & Treatments | |
| Pests & Treatment | |

**Notes:**

# Cherry Files

| | |
|---|---|
| Size | |
| Fruit Qualities | |
| Bloom Time | |
| Cross-Pollinator | |
| Harvest Time | |
| Source | |
| Disease Resistance | |
| Pruning | |
| Soil Amendments | |
| Diseases & Treatments | |
| Pests & Treatment | |

**Notes:**

# Cherry Files

| | |
|---|---|
| Size | |
| Fruit Qualities | |
| Bloom Time | |
| Cross-Pollinator | |
| Harvest Time | |
| Source | |
| Disease Resistance | |
| Pruning | |
| Soil Amendments | |
| Diseases & Treatments | |
| Pests & Treatment | |

**Notes:**

# Cherry Files

| | |
|---|---|
| Size | |
| Fruit Qualities | |
| Bloom Time | |
| Cross-Pollinator | |
| Harvest Time | |
| Source | |
| Disease Resistance | |
| Pruning | |
| Soil Amendments | |
| Diseases & Treatments | |
| Pests & Treatment | |

**Notes:**

# Cherry Files

| | |
|---|---|
| Size | |
| Fruit Qualities | |
| Bloom Time | |
| Cross-Pollinator | |
| Harvest Time | |
| Source | |
| Disease Resistance | |
| Pruning | |
| Soil Amendments | |
| Diseases & Treatments | |
| Pests & Treatment | |

**Notes:**

# Cherry Files

| | |
|---|---|
| Size | |
| Fruit Qualities | |
| Bloom Time | |
| Cross-Pollinator | |
| Harvest Time | |
| Source | |
| Disease Resistance | |
| Pruning | |
| Soil Amendments | |
| Diseases & Treatments | |
| Pests & Treatment | |

**Notes:**

# Cherry Files

| | |
|---|---|
| Size | |
| Fruit Qualities | |
| Bloom Time | |
| Cross-Pollinator | |
| Harvest Time | |
| Source | |
| Disease Resistance | |
| Pruning | |
| Soil Amendments | |
| Diseases & Treatments | |
| Pests & Treatment | |

**Notes:**

# Nectarine Files

| | |
|---|---|
| Size | |
| Fruit Qualities | |
| Bloom Time | |
| Cross-Pollinator | |
| Harvest Time | |
| Source | |
| Disease Resistance | |
| Pruning | |
| Soil Amendments | |
| Diseases & Treatments | |
| Pests & Treatment | |

**Notes:**

# Nectarine Files

______________________________

| | |
|---|---|
| Size | |
| Fruit Qualities | |
| Bloom Time | |
| Cross-Pollinator | |
| Harvest Time | |
| Source | |
| Disease Resistance | |
| Pruning | |
| Soil Amendments | |
| Diseases & Treatments | |
| Pests & Treatment | |

**Notes:**

# Peach Files

| | |
|---|---|
| Size | |
| Fruit Qualities | |
| Bloom Time | |
| Cross-Pollinator | |
| Harvest Time | |
| Source | |
| Disease Resistance | |
| Pruning | |
| Soil Amendments | |
| Diseases & Treatments | |
| Pests & Treatment | |

**Notes:**

# Peach Files

_______________

| | |
|---|---|
| Size | |
| Fruit Qualities | |
| Bloom Time | |
| Cross-Pollinator | |
| Harvest Time | |
| Source | |
| Disease Resistance | |
| Pruning | |
| Soil Amendments | |
| Diseases & Treatments | |
| Pests & Treatment | |

**Notes:**

# Peach Files

| | |
|---|---|
| Size | |
| Fruit Qualities | |
| Bloom Time | |
| Cross-Pollinator | |
| Harvest Time | |
| Source | |
| Disease Resistance | |
| Pruning | |
| Soil Amendments | |
| Diseases & Treatments | |
| Pests & Treatment | |

**Notes:**

# Peach Files

| | |
|---|---|
| Size | |
| Fruit Qualities | |
| Bloom Time | |
| Cross-Pollinator | |
| Harvest Time | |
| Source | |
| Disease Resistance | |
| Pruning | |
| Soil Amendments | |
| Diseases & Treatments | |
| Pests & Treatment | |

**Notes:**

# Peach Files

| | |
|---|---|
| Size | |
| Fruit Qualities | |
| Bloom Time | |
| Cross-Pollinator | |
| Harvest Time | |
| Source | |
| Disease Resistance | |
| Pruning | |
| Soil Amendments | |
| Diseases & Treatments | |
| Pests & Treatment | |

**Notes:**

# Peach Files

| | |
|---|---|
| Size | |
| Fruit Qualities | |
| Bloom Time | |
| Cross-Pollinator | |
| Harvest Time | |
| Source | |
| Disease Resistance | |
| Pruning | |
| Soil Amendments | |
| Diseases & Treatments | |
| Pests & Treatment | |

**Notes:**

# Peach Files

| | |
|---|---|
| Size | |
| Fruit Qualities | |
| Bloom Time | |
| Cross-Pollinator | |
| Harvest Time | |
| Source | |
| Disease Resistance | |
| Pruning | |
| Soil Amendments | |
| Diseases & Treatments | |
| Pests & Treatment | |

**Notes:**

# Peach Files

| | |
|---|---|
| Size | |
| Fruit Qualities | |
| Bloom Time | |
| Cross-Pollinator | |
| Harvest Time | |
| Source | |
| Disease Resistance | |
| Pruning | |
| Soil Amendments | |
| Diseases & Treatments | |
| Pests & Treatment | |

**Notes:**

# Plum Files

| | |
|---|---|
| Size | |
| Fruit Qualities | |
| Bloom Time | |
| Cross-Pollinator | |
| Harvest Time | |
| Source | |
| Disease Resistance | |
| Pruning | |
| Soil Amendments | |
| Diseases & Treatments | |
| Pests & Treatment | |

**Notes:**

# Plum Files

_______________

| | |
|---|---|
| Size | |
| Fruit Qualities | |
| Bloom Time | |
| Cross-Pollinator | |
| Harvest Time | |
| Source | |
| Disease Resistance | |
| Pruning | |
| Soil Amendments | |
| Diseases & Treatments | |
| Pests & Treatment | |

**Notes:**

# Plum Files

| | |
|---|---|
| Size | |
| Fruit Qualities | |
| Bloom Time | |
| Cross-Pollinator | |
| Harvest Time | |
| Source | |
| Disease Resistance | |
| Pruning | |
| Soil Amendments | |
| Diseases & Treatments | |
| Pests & Treatment | |

**Notes:**

# Plum Files

| | |
|---|---|
| Size | |
| Fruit Qualities | |
| Bloom Time | |
| Cross-Pollinator | |
| Harvest Time | |
| Source | |
| Disease Resistance | |
| Pruning | |
| Soil Amendments | |
| Diseases & Treatments | |
| Pests & Treatment | |

**Notes:**

# Plum Files

______________________________

| | |
|---|---|
| Size | |
| Fruit Qualities | |
| Bloom Time | |
| Cross-Pollinator | |
| Harvest Time | |
| Source | |
| Disease Resistance | |
| Pruning | |
| Soil Amendments | |
| Diseases & Treatments | |
| Pests & Treatment | |

**Notes:**

# Plum Files

| | |
|---|---|
| Size | |
| Fruit Qualities | |
| Bloom Time | |
| Cross-Pollinator | |
| Harvest Time | |
| Source | |
| Disease Resistance | |
| Pruning | |
| Soil Amendments | |
| Diseases & Treatments | |
| Pests & Treatment | |

**Notes:**

# Plum Files

______________________

| | |
|---|---|
| Size | |
| Fruit Qualities | |
| Bloom Time | |
| Cross-Pollinator | |
| Harvest Time | |
| Source | |
| Disease Resistance | |
| Pruning | |
| Soil Amendments | |
| Diseases & Treatments | |
| Pests & Treatment | |

**Notes:**

# Plum Files

| | |
|---|---|
| Size | |
| Fruit Qualities | |
| Bloom Time | |
| Cross-Pollinator | |
| Harvest Time | |
| Source | |
| Disease Resistance | |
| Pruning | |
| Soil Amendments | |
| Diseases & Treatments | |
| Pests & Treatment | |

**Notes:**

 

## Notes

 

# Citrus

# Grapefruit Files

| | |
|---|---|
| Size | |
| Fruit Qualities | |
| Bloom Time | |
| Cross-Pollinator | |
| Harvest Time | |
| Source | |
| Disease Resistance | |
| Pruning | |
| Soil Amendments | |
| Diseases & Treatments | |
| Pests & Treatment | |

**Notes:**

# Grapefruit Files

| | |
|---|---|
| Size | |
| Fruit Qualities | |
| Bloom Time | |
| Cross-Pollinator | |
| Harvest Time | |
| Source | |
| Disease Resistance | |
| Pruning | |
| Soil Amendments | |
| Diseases & Treatments | |
| Pests & Treatment | |

**Notes:**

# Lemon Files

| | |
|---|---|
| Size | |
| Fruit Qualities | |
| Bloom Time | |
| Cross-Pollinator | |
| Harvest Time | |
| Source | |
| Disease Resistance | |
| Pruning | |
| Soil Amendments | |
| Diseases & Treatments | |
| Pests & Treatment | |

**Notes:**

# Lemon Files

______________________________

| | |
|---|---|
| Size | |
| Fruit Qualities | |
| Bloom Time | |
| Cross-Pollinator | |
| Harvest Time | |
| Source | |
| Disease Resistance | |
| Pruning | |
| Soil Amendments | |
| Diseases & Treatments | |
| Pests & Treatment | |

**Notes:**

# Lemon Files

| | |
|---|---|
| Size | |
| Fruit Qualities | |
| Bloom Time | |
| Cross-Pollinator | |
| Harvest Time | |
| Source | |
| Disease Resistance | |
| Pruning | |
| Soil Amendments | |
| Diseases & Treatments | |
| Pests & Treatment | |

**Notes:**

# Lemon Files

| | |
|---|---|
| Size | |
| Fruit Qualities | |
| Bloom Time | |
| Cross-Pollinator | |
| Harvest Time | |
| Source | |
| Disease Resistance | |
| Pruning | |
| Soil Amendments | |
| Diseases & Treatments | |
| Pests & Treatment | |

**Notes:**

# Lime Files

| | |
|---|---|
| Size | |
| Fruit Qualities | |
| Bloom Time | |
| Cross-Pollinator | |
| Harvest Time | |
| Source | |
| Disease Resistance | |
| Pruning | |
| Soil Amendments | |
| Diseases & Treatments | |
| Pests & Treatment | |

**Notes:**

# Lime Files

______________________

| | |
|---|---|
| Size | |
| Fruit Qualities | |
| Bloom Time | |
| Cross-Pollinator | |
| Harvest Time | |
| Source | |
| Disease Resistance | |
| Pruning | |
| Soil Amendments | |
| Diseases & Treatments | |
| Pests & Treatment | |

**Notes:**

# Orange Files

| | |
|---|---|
| Size | |
| Fruit Qualities | |
| Bloom Time | |
| Cross-Pollinator | |
| Harvest Time | |
| Source | |
| Disease Resistance | |
| Pruning | |
| Soil Amendments | |
| Diseases & Treatments | |
| Pests & Treatment | |

**Notes:**

# Orange Files

| | |
|---|---|
| Size | |
| Fruit Qualities | |
| Bloom Time | |
| Cross-Pollinator | |
| Harvest Time | |
| Source | |
| Disease Resistance | |
| Pruning | |
| Soil Amendments | |
| Diseases & Treatments | |
| Pests & Treatment | |

**Notes:**

# Orange Files

| | |
|---|---|
| Size | |
| Fruit Qualities | |
| Bloom Time | |
| Cross-Pollinator | |
| Harvest Time | |
| Source | |
| Disease Resistance | |
| Pruning | |
| Soil Amendments | |
| Diseases & Treatments | |
| Pests & Treatment | |

**Notes:**

# Orange Files

| | |
|---|---|
| Size | |
| Fruit Qualities | |
| Bloom Time | |
| Cross-Pollinator | |
| Harvest Time | |
| Source | |
| Disease Resistance | |
| Pruning | |
| Soil Amendments | |
| Diseases & Treatments | |
| Pests & Treatment | |

**Notes:**

# Citrus Files

| | |
|---|---|
| Size | |
| Fruit Qualities | |
| Bloom Time | |
| Cross-Pollinator | |
| Harvest Time | |
| Source | |
| Disease Resistance | |
| Pruning | |
| Soil Amendments | |
| Diseases & Treatments | |
| Pests & Treatment | |

**Notes:**

# Citrus Files

| | |
|---|---|
| Size | |
| Fruit Qualities | |
| Bloom Time | |
| Cross-Pollinator | |
| Harvest Time | |
| Source | |
| Disease Resistance | |
| Pruning | |
| Soil Amendments | |
| Diseases & Treatments | |
| Pests & Treatment | |

**Notes:**

 

# Notes

 

# Berries

# Blackberry Files

| | |
|---|---|
| Thorns | |
| Fruit Qualities | |
| Bloom Time | |
| Cross-Pollinator | |
| Harvest Time | |
| Source | |
| Disease Resistance | |
| Pruning | |
| Soil Amendments | |
| Diseases & Treatments | |
| Pests & Treatment | |

**Notes:**

# Blackberry Files

| | |
|---|---|
| Thorns | |
| Fruit Qualities | |
| Bloom Time | |
| Cross-Pollinator | |
| Harvest Time | |
| Source | |
| Disease Resistance | |
| Pruning | |
| Soil Amendments | |
| Diseases & Treatments | |
| Pests & Treatment | |

**Notes:**

# Blackberry Files

| | |
|---|---|
| Thorns | |
| Fruit Qualities | |
| Bloom Time | |
| Cross-Pollinator | |
| Harvest Time | |
| Source | |
| Disease Resistance | |
| Pruning | |
| Soil Amendments | |
| Diseases & Treatments | |
| Pests & Treatment | |

**Notes:**

# Blackberry Files

| | |
|---|---|
| Thorns | |
| Fruit Qualities | |
| Bloom Time | |
| Cross-Pollinator | |
| Harvest Time | |
| Source | |
| Disease Resistance | |
| Pruning | |
| Soil Amendments | |
| Diseases & Treatments | |
| Pests & Treatment | |

**Notes:**

# Blueberry Files

| | |
|---|---|
| Type | |
| Fruit Qualities | |
| Bloom Time | |
| Cross-Pollinator | |
| Harvest Time | |
| Source | |
| Disease Resistance | |
| Pruning | |
| Soil Amendments | |
| Diseases & Treatments | |
| Pests & Treatment | |

**Notes:**

# Blueberry Files

| | |
|---|---|
| Type | |
| Fruit Qualities | |
| Bloom Time | |
| Cross-Pollinator | |
| Harvest Time | |
| Source | |
| Disease Resistance | |
| Pruning | |
| Soil Amendments | |
| Diseases & Treatments | |
| Pests & Treatment | |

**Notes:**

# Blueberry Files

| | |
|---|---|
| Type | |
| Fruit Qualities | |
| Bloom Time | |
| Cross-Pollinator | |
| Harvest Time | |
| Source | |
| Disease Resistance | |
| Pruning | |
| Soil Amendments | |
| Diseases & Treatments | |
| Pests & Treatment | |

**Notes:**

# Blueberry Files

| | |
|---|---|
| Type | |
| Fruit Qualities | |
| Bloom Time | |
| Cross-Pollinator | |
| Harvest Time | |
| Source | |
| Disease Resistance | |
| Pruning | |
| Soil Amendments | |
| Diseases & Treatments | |
| Pests & Treatment | |

**Notes:**

# Blueberry Files

| | |
|---|---|
| Type | |
| Fruit Qualities | |
| Bloom Time | |
| Cross-Pollinator | |
| Harvest Time | |
| Source | |
| Disease Resistance | |
| Pruning | |
| Soil Amendments | |
| Diseases & Treatments | |
| Pests & Treatment | |

**Notes:**

# Blueberry Files

| | |
|---|---|
| Type | |
| Fruit Qualities | |
| Bloom Time | |
| Cross-Pollinator | |
| Harvest Time | |
| Source | |
| Disease Resistance | |
| Pruning | |
| Soil Amendments | |
| Diseases & Treatments | |
| Pests & Treatment | |

**Notes:**

# Cranberry Files

| | |
|---|---|
| Type | |
| Fruit Qualities | |
| Bloom Time | |
| Cross-Pollinator | |
| Harvest Time | |
| Source | |
| Disease Resistance | |
| Pruning | |
| Soil Amendments | |
| Diseases & Treatments | |
| Pests & Treatment | |

**Notes:**

# Cranberry Files

| | |
|---|---|
| Type | |
| Fruit Qualities | |
| Bloom Time | |
| Cross-Pollinator | |
| Harvest Time | |
| Source | |
| Disease Resistance | |
| Pruning | |
| Soil Amendments | |
| Diseases & Treatments | |
| Pests & Treatment | |

**Notes:**

# Currant Files

| | |
|---|---|
| Type | |
| Fruit Qualities | |
| Bloom Time | |
| Cross-Pollinator | |
| Harvest Time | |
| Source | |
| Disease Resistance | |
| Pruning | |
| Soil Amendments | |
| Diseases & Treatments | |
| Pests & Treatment | |

**Notes:**

# Currant Files

| | |
|---|---|
| Type | |
| Fruit Qualities | |
| Bloom Time | |
| Cross-Pollinator | |
| Harvest Time | |
| Source | |
| Disease Resistance | |
| Pruning | |
| Soil Amendments | |
| Diseases & Treatments | |
| Pests & Treatment | |

**Notes:**

# Elderberry Files

| | |
|---|---|
| Type | |
| Fruit Qualities | |
| Bloom Time | |
| Cross-Pollinator | |
| Harvest Time | |
| Source | |
| Disease Resistance | |
| Pruning | |
| Soil Amendments | |
| Diseases & Treatments | |
| Pests & Treatment | |

**Notes:**

# Elderberry Files

| | |
|---|---|
| Type | |
| Fruit Qualities | |
| Bloom Time | |
| Cross-Pollinator | |
| Harvest Time | |
| Source | |
| Disease Resistance | |
| Pruning | |
| Soil Amendments | |
| Diseases & Treatments | |
| Pests & Treatment | |

**Notes:**

# Gooseberry Files

| | |
|---|---|
| Type | |
| Fruit Qualities | |
| Bloom Time | |
| Cross-Pollinator | |
| Harvest Time | |
| Source | |
| Disease Resistance | |
| Pruning | |
| Soil Amendments | |
| Diseases & Treatments | |
| Pests & Treatment | |

**Notes:**

# Gooseberry Files

| | |
|---|---|
| Type | |
| Fruit Qualities | |
| Bloom Time | |
| Cross-Pollinator | |
| Harvest Time | |
| Source | |
| Disease Resistance | |
| Pruning | |
| Soil Amendments | |
| Diseases & Treatments | |
| Pests & Treatment | |

**Notes:**

# Huckleberry Files

| | |
|---|---|
| Type | |
| Fruit Qualities | |
| Bloom Time | |
| Cross-Pollinator | |
| Harvest Time | |
| Source | |
| Disease Resistance | |
| Pruning | |
| Soil Amendments | |
| Diseases & Treatments | |
| Pests & Treatment | |

**Notes:**

# Huckleberry Files

| | |
|---|---|
| Type | |
| Fruit Qualities | |
| Bloom Time | |
| Cross-Pollinator | |
| Harvest Time | |
| Source | |
| Disease Resistance | |
| Pruning | |
| Soil Amendments | |
| Diseases & Treatments | |
| Pests & Treatment | |

**Notes:**

# Mulberry Files

| | |
|---|---|
| Type | |
| Fruit Qualities | |
| Bloom Time | |
| Cross-Pollinator | |
| Harvest Time | |
| Source | |
| Disease Resistance | |
| Pruning | |
| Soil Amendments | |
| Diseases & Treatments | |
| Pests & Treatment | |

**Notes:**

# Mulberry Files

| | |
|---|---|
| Type | |
| Fruit Qualities | |
| Bloom Time | |
| Cross-Pollinator | |
| Harvest Time | |
| Source | |
| Disease Resistance | |
| Pruning | |
| Soil Amendments | |
| Diseases & Treatments | |
| Pests & Treatment | |

**Notes:**

# Raspberry Files

| | |
|---|---|
| Thorns | |
| Fruit Qualities | |
| Bloom Time | |
| Cross-Pollinator | |
| Harvest Time | |
| Source | |
| Disease Resistance | |
| Pruning | |
| Soil Amendments | |
| Diseases & Treatments | |
| Pests & Treatment | |

**Notes:**

# Raspberry Files

| | |
|---|---|
| Thorns | |
| Fruit Qualities | |
| Bloom Time | |
| Cross-Pollinator | |
| Harvest Time | |
| Source | |
| Disease Resistance | |
| Pruning | |
| Soil Amendments | |
| Diseases & Treatments | |
| Pests & Treatment | |

**Notes:**

# Raspberry Files

| | |
|---|---|
| Thorns | |
| Fruit Qualities | |
| Bloom Time | |
| Cross-Pollinator | |
| Harvest Time | |
| Source | |
| Disease Resistance | |
| Pruning | |
| Soil Amendments | |
| Diseases & Treatments | |
| Pests & Treatment | |

**Notes:**

# Raspberry Files

| | |
|---|---|
| Thorns | |
| Fruit Qualities | |
| Bloom Time | |
| Cross-Pollinator | |
| Harvest Time | |
| Source | |
| Disease Resistance | |
| Pruning | |
| Soil Amendments | |
| Diseases & Treatments | |
| Pests & Treatment | |

**Notes:**

# Raspberry Files

| | |
|---|---|
| Thorns | |
| Fruit Qualities | |
| Bloom Time | |
| Cross-Pollinator | |
| Harvest Time | |
| Source | |
| Disease Resistance | |
| Pruning | |
| Soil Amendments | |
| Diseases & Treatments | |
| Pests & Treatment | |

**Notes:**

# Raspberry Files

| | |
|---|---|
| Thorns | |
| Fruit Qualities | |
| Bloom Time | |
| Cross-Pollinator | |
| Harvest Time | |
| Source | |
| Disease Resistance | |
| Pruning | |
| Soil Amendments | |
| Diseases & Treatments | |
| Pests & Treatment | |

**Notes:**

# Strawberry Files

| | |
|---|---|
| Type | |
| Fruit Qualities | |
| Bloom Time | |
| Cross-Pollinator | |
| Harvest Time | |
| Source | |
| Disease Resistance | |
| Pruning | |
| Soil Amendments | |
| Diseases & Treatments | |
| Pests & Treatment | |

**Notes:**

# Strawberry Files

| | |
|---|---|
| Type | |
| Fruit Qualities | |
| Bloom Time | |
| Cross-Pollinator | |
| Harvest Time | |
| Source | |
| Disease Resistance | |
| Pruning | |
| Soil Amendments | |
| Diseases & Treatments | |
| Pests & Treatment | |

**Notes:**

# Strawberry Files

| | |
|---|---|
| Type | |
| Fruit Qualities | |
| Bloom Time | |
| Cross-Pollinator | |
| Harvest Time | |
| Source | |
| Disease Resistance | |
| Pruning | |
| Soil Amendments | |
| Diseases & Treatments | |
| Pests & Treatment | |

**Notes:**

# Strawberry Files

| | |
|---|---|
| Type | |
| Fruit Qualities | |
| Bloom Time | |
| Cross-Pollinator | |
| Harvest Time | |
| Source | |
| Disease Resistance | |
| Pruning | |
| Soil Amendments | |
| Diseases & Treatments | |
| Pests & Treatment | |

**Notes:**

# Strawberry Files

| | |
|---|---|
| Type | |
| Fruit Qualities | |
| Bloom Time | |
| Cross-Pollinator | |
| Harvest Time | |
| Source | |
| Disease Resistance | |
| Pruning | |
| Soil Amendments | |
| Diseases & Treatments | |
| Pests & Treatment | |

**Notes:**

# Strawberry Files

| | |
|---|---|
| Type | |
| Fruit Qualities | |
| Bloom Time | |
| Cross-Pollinator | |
| Harvest Time | |
| Source | |
| Disease Resistance | |
| Pruning | |
| Soil Amendments | |
| Diseases & Treatments | |
| Pests & Treatment | |

**Notes:**

# Berry Files

| | |
|---|---|
| Type | |
| Fruit Qualities | |
| Bloom Time | |
| Cross-Pollinator | |
| Harvest Time | |
| Source | |
| Disease Resistance | |
| Pruning | |
| Soil Amendments | |
| Diseases & Treatments | |
| Pests & Treatment | |

**Notes:**

# Berry Files

| | |
|---|---|
| Type | |
| Fruit Qualities | |
| Bloom Time | |
| Cross-Pollinator | |
| Harvest Time | |
| Source | |
| Disease Resistance | |
| Pruning | |
| Soil Amendments | |
| Diseases & Treatments | |
| Pests & Treatment | |

**Notes:**

# Berry Files

| | |
|---|---|
| Type | |
| Fruit Qualities | |
| Bloom Time | |
| Cross-Pollinator | |
| Harvest Time | |
| Source | |
| Disease Resistance | |
| Pruning | |
| Soil Amendments | |
| Diseases & Treatments | |
| Pests & Treatment | |

**Notes:**

# Berry Files

_______________

| | |
|---|---|
| Type | |
| Fruit Qualities | |
| Bloom Time | |
| Cross-Pollinator | |
| Harvest Time | |
| Source | |
| Disease Resistance | |
| Pruning | |
| Soil Amendments | |
| Diseases & Treatments | |
| Pests & Treatment | |

**Notes:**

 

# Notes

 

# Vines

# Grape Files

| | |
|---|---|
| Type | |
| Fruit Qualities | |
| Bloom Time | |
| Cross-Pollinator | |
| Harvest Time | |
| Source | |
| Disease Resistance | |
| Pruning | |
| Soil Amendments | |
| Diseases & Treatments | |
| Pests & Treatment | |

**Notes:**

# Grape Files

| | |
|---|---|
| Type | |
| Fruit Qualities | |
| Bloom Time | |
| Cross-Pollinator | |
| Harvest Time | |
| Source | |
| Disease Resistance | |
| Pruning | |
| Soil Amendments | |
| Diseases & Treatments | |
| Pests & Treatment | |

**Notes:**

# Grape Files

| | |
|---|---|
| Type | |
| Fruit Qualities | |
| Bloom Time | |
| Cross-Pollinator | |
| Harvest Time | |
| Source | |
| Disease Resistance | |
| Pruning | |
| Soil Amendments | |
| Diseases & Treatments | |
| Pests & Treatment | |

**Notes:**

# Grape Files

| | |
|---|---|
| Type | |
| Fruit Qualities | |
| Bloom Time | |
| Cross-Pollinator | |
| Harvest Time | |
| Source | |
| Disease Resistance | |
| Pruning | |
| Soil Amendments | |
| Diseases & Treatments | |
| Pests & Treatment | |

**Notes:**

# Grape Files

| | |
|---|---|
| Type | |
| Fruit Qualities | |
| Bloom Time | |
| Cross-Pollinator | |
| Harvest Time | |
| Source | |
| Disease Resistance | |
| Pruning | |
| Soil Amendments | |
| Diseases & Treatments | |
| Pests & Treatment | |

**Notes:**

# Grape Files

| | |
|---|---|
| Type | |
| Fruit Qualities | |
| Bloom Time | |
| Cross-Pollinator | |
| Harvest Time | |
| Source | |
| Disease Resistance | |
| Pruning | |
| Soil Amendments | |
| Diseases & Treatments | |
| Pests & Treatment | |

**Notes:**

# Grape Files

| | |
|---|---|
| Type | |
| Fruit Qualities | |
| Bloom Time | |
| Cross-Pollinator | |
| Harvest Time | |
| Source | |
| Disease Resistance | |
| Pruning | |
| Soil Amendments | |
| Diseases & Treatments | |
| Pests & Treatment | |

**Notes:**

# Grape Files

| | |
|---|---|
| Type | |
| Fruit Qualities | |
| Bloom Time | |
| Cross-Pollinator | |
| Harvest Time | |
| Source | |
| Disease Resistance | |
| Pruning | |
| Soil Amendments | |
| Diseases & Treatments | |
| Pests & Treatment | |

**Notes:**

# Grape Files

| | |
|---|---|
| Type | |
| Fruit Qualities | |
| Bloom Time | |
| Cross-Pollinator | |
| Harvest Time | |
| Source | |
| Disease Resistance | |
| Pruning | |
| Soil Amendments | |
| Diseases & Treatments | |
| Pests & Treatment | |

**Notes:**

# Grape Files

| | |
|---|---|
| Type | |
| Fruit Qualities | |
| Bloom Time | |
| Cross-Pollinator | |
| Harvest Time | |
| Source | |
| Disease Resistance | |
| Pruning | |
| Soil Amendments | |
| Diseases & Treatments | |
| Pests & Treatment | |

**Notes:**

# Kiwi Files

| | |
|---|---|
| Type | |
| Fruit Qualities | |
| Bloom Time | |
| Cross-Pollinator | |
| Harvest Time | |
| Source | |
| Disease Resistance | |
| Pruning | |
| Soil Amendments | |
| Diseases & Treatments | |
| Pests & Treatment | |

**Notes:**

# Kiwi Files

| | |
|---|---|
| Type | |
| Fruit Qualities | |
| Bloom Time | |
| Cross-Pollinator | |
| Harvest Time | |
| Source | |
| Disease Resistance | |
| Pruning | |
| Soil Amendments | |
| Diseases & Treatments | |
| Pests & Treatment | |

**Notes:**

# Passionfruit Files

| | |
|---|---|
| Type | |
| Fruit Qualities | |
| Bloom Time | |
| Cross-Pollinator | |
| Harvest Time | |
| Source | |
| Disease Resistance | |
| Pruning | |
| Soil Amendments | |
| Diseases & Treatments | |
| Pests & Treatment | |

**Notes:**

# Passionfruit Files

| | |
|---|---|
| Type | |
| Fruit Qualities | |
| Bloom Time | |
| Cross-Pollinator | |
| Harvest Time | |
| Source | |
| Disease Resistance | |
| Pruning | |
| Soil Amendments | |
| Diseases & Treatments | |
| Pests & Treatment | |

**Notes:**

 

# Notes

 

# Nuts

# Almond Files

| | |
|---|---|
| Type | |
| Nut Qualities | |
| Bloom Time | |
| Cross-Pollinator | |
| Harvest Time | |
| Source | |
| Disease Resistance | |
| Pruning | |
| Soil Amendments | |
| Diseases & Treatments | |
| Pests & Treatment | |

**Notes:**

# Almond Files

| | |
|---|---|
| Type | |
| Nut Qualities | |
| Bloom Time | |
| Cross-Pollinator | |
| Harvest Time | |
| Source | |
| Disease Resistance | |
| Pruning | |
| Soil Amendments | |
| Diseases & Treatments | |
| Pests & Treatment | |

**Notes:**

# Cashew Files

| | |
|---|---|
| Type | |
| Nut Qualities | |
| Bloom Time | |
| Cross-Pollinator | |
| Harvest Time | |
| Source | |
| Disease Resistance | |
| Pruning | |
| Soil Amendments | |
| Diseases & Treatments | |
| Pests & Treatment | |

**Notes:**

# Cashew Files

| | |
|---|---|
| Type | |
| Nut Qualities | |
| Bloom Time | |
| Cross-Pollinator | |
| Harvest Time | |
| Source | |
| Disease Resistance | |
| Pruning | |
| Soil Amendments | |
| Diseases & Treatments | |
| Pests & Treatment | |

**Notes:**

# Chestnut Files

| | |
|---|---|
| Type | |
| Nut Qualities | |
| Bloom Time | |
| Cross-Pollinator | |
| Harvest Time | |
| Source | |
| Disease Resistance | |
| Pruning | |
| Soil Amendments | |
| Diseases & Treatments | |
| Pests & Treatment | |

**Notes:**

# Chestnut Files

| | |
|---|---|
| Type | |
| Nut Qualities | |
| Bloom Time | |
| Cross-Pollinator | |
| Harvest Time | |
| Source | |
| Disease Resistance | |
| Pruning | |
| Soil Amendments | |
| Diseases & Treatments | |
| Pests & Treatment | |

**Notes:**

# Hazelnut Files

| | |
|---|---|
| Type | |
| Nut Qualities | |
| Bloom Time | |
| Cross-Pollinator | |
| Harvest Time | |
| Source | |
| Disease Resistance | |
| Pruning | |
| Soil Amendments | |
| Diseases & Treatments | |
| Pests & Treatment | |

**Notes:**

# Hazelnut Files

| | |
|---|---|
| Type | |
| Nut Qualities | |
| Bloom Time | |
| Cross-Pollinator | |
| Harvest Time | |
| Source | |
| Disease Resistance | |
| Pruning | |
| Soil Amendments | |
| Diseases & Treatments | |
| Pests & Treatment | |

**Notes:**

# Pecan Files

| | |
|---|---|
| Type | |
| Nut Qualities | |
| Bloom Time | |
| Cross-Pollinator | |
| Harvest Time | |
| Source | |
| Disease Resistance | |
| Pruning | |
| Soil Amendments | |
| Diseases & Treatments | |
| Pests & Treatment | |

**Notes:**

# Pecan Files

| | |
|---|---|
| Type | |
| Nut Qualities | |
| Bloom Time | |
| Cross-Pollinator | |
| Harvest Time | |
| Source | |
| Disease Resistance | |
| Pruning | |
| Soil Amendments | |
| Diseases & Treatments | |
| Pests & Treatment | |

**Notes:**

# Pistachio Files

| | |
|---|---|
| Type | |
| Nut Qualities | |
| Bloom Time | |
| Cross-Pollinator | |
| Harvest Time | |
| Source | |
| Disease Resistance | |
| Pruning | |
| Soil Amendments | |
| Diseases & Treatments | |
| Pests & Treatment | |

**Notes:**

# Postachio Files

| | |
|---|---|
| Type | |
| Nut Qualities | |
| Bloom Time | |
| Cross-Pollinator | |
| Harvest Time | |
| Source | |
| Disease Resistance | |
| Pruning | |
| Soil Amendments | |
| Diseases & Treatments | |
| Pests & Treatment | |

**Notes:**

# *Walnut Files*

| | |
|---|---|
| Type | |
| Nut Qualities | |
| Bloom Time | |
| Cross-Pollinator | |
| Harvest Time | |
| Source | |
| Disease Resistance | |
| Pruning | |
| Soil Amendments | |
| Diseases & Treatments | |
| Pests & Treatment | |

**Notes:**

# Walnut Files

| | |
|---|---|
| Type | |
| Nut Qualities | |
| Bloom Time | |
| Cross-Pollinator | |
| Harvest Time | |
| Source | |
| Disease Resistance | |
| Pruning | |
| Soil Amendments | |
| Diseases & Treatments | |
| Pests & Treatment | |

**Notes:**

 

# Notes

 

# Natives Wild Fruits

# Persimmon Files

| | |
|---|---|
| Type | |
| Fruit Qualities | |
| Bloom Time | |
| Cross-Pollinator | |
| Harvest Time | |
| Source | |
| Disease Resistance | |
| Pruning | |
| Soil Amendments | |
| Diseases & Treatments | |
| Pests & Treatment | |

**Notes:**

# Persimmon Files

| | |
|---|---|
| Type | |
| Fruit Qualities | |
| Bloom Time | |
| Cross-Pollinator | |
| Harvest Time | |
| Source | |
| Disease Resistance | |
| Pruning | |
| Soil Amendments | |
| Diseases & Treatments | |
| Pests & Treatment | |

**Notes:**

# Pawpaw Files

| | |
|---|---|
| Type | |
| Fruit Qualities | |
| Bloom Time | |
| Cross-Pollinator | |
| Harvest Time | |
| Source | |
| Disease Resistance | |
| Pruning | |
| Soil Amendments | |
| Diseases & Treatments | |
| Pests & Treatment | |

**Notes:**

# Crabapple Files

| | |
|---|---|
| Type | |
| Fruit Qualities | |
| Bloom Time | |
| Cross-Pollinator | |
| Harvest Time | |
| Source | |
| Disease Resistance | |
| Pruning | |
| Soil Amendments | |
| Diseases & Treatments | |
| Pests & Treatment | |

**Notes:**

# Wild Plum Files

| | |
|---|---|
| Type | |
| Fruit Qualities | |
| Bloom Time | |
| Cross-Pollinator | |
| Harvest Time | |
| Source | |
| Disease Resistance | |
| Pruning | |
| Soil Amendments | |
| Diseases & Treatments | |
| Pests & Treatment | |

**Notes:**

# Prickly Pear Files

| | |
|---|---|
| Type | |
| Fruit Qualities | |
| Bloom Time | |
| Cross-Pollinator | |
| Harvest Time | |
| Source | |
| Disease Resistance | |
| Pruning | |
| Soil Amendments | |
| Diseases & Treatments | |
| Pests & Treatment | |

**Notes:**

# Acai Berry Files

| | |
|---|---|
| Type | |
| Fruit Qualities | |
| Bloom Time | |
| Cross-Pollinator | |
| Harvest Time | |
| Source | |
| Disease Resistance | |
| Pruning | |
| Soil Amendments | |
| Diseases & Treatments | |
| Pests & Treatment | |

**Notes:**

# Lingonberry Files

| | |
|---|---|
| Type | |
| Fruit Qualities | |
| Bloom Time | |
| Cross-Pollinator | |
| Harvest Time | |
| Source | |
| Disease Resistance | |
| Pruning | |
| Soil Amendments | |
| Diseases & Treatments | |
| Pests & Treatment | |

**Notes:**

# Rose Hip Files

| | |
|---|---|
| Type | |
| Fruit Qualities | |
| Bloom Time | |
| Cross-Pollinator | |
| Harvest Time | |
| Source | |
| Disease Resistance | |
| Pruning | |
| Soil Amendments | |
| Diseases & Treatments | |
| Pests & Treatment | |

**Notes:**

# Wild Blueberry Files

| | |
|---|---|
| Type | |
| Fruit Qualities | |
| Bloom Time | |
| Cross-Pollinator | |
| Harvest Time | |
| Source | |
| Disease Resistance | |
| Pruning | |
| Soil Amendments | |
| Diseases & Treatments | |
| Pests & Treatment | |

**Notes:**

# *Wild Blackberry Files*

______________________

| | |
|---|---|
| Type | |
| Fruit Qualities | |
| Bloom Time | |
| Cross-Pollinator | |
| Harvest Time | |
| Source | |
| Disease Resistance | |
| Pruning | |
| Soil Amendments | |
| Diseases & Treatments | |
| Pests & Treatment | |

**Notes:**

# Wild Raspberry Files

______________________

| | |
|---|---|
| Type | |
| Fruit Qualities | |
| Bloom Time | |
| Cross-Pollinator | |
| Harvest Time | |
| Source | |
| Disease Resistance | |
| Pruning | |
| Soil Amendments | |
| Diseases & Treatments | |
| Pests & Treatment | |

**Notes:**

 

# Wild Strawberry Files

______________________

| | |
|---|---|
| Type | |
| Fruit Qualities | |
| Bloom Time | |
| Cross-Pollinator | |
| Harvest Time | |
| Source | |
| Disease Resistance | |
| Pruning | |
| Soil Amendments | |
| Diseases & Treatments | |
| Pests & Treatment | |

**Notes:**

# Muscadine Files

| | |
|---|---|
| Type | |
| Fruit Qualities | |
| Bloom Time | |
| Cross-Pollinator | |
| Harvest Time | |
| Source | |
| Disease Resistance | |
| Pruning | |
| Soil Amendments | |
| Diseases & Treatments | |
| Pests & Treatment | |

**Notes:**

# Chokecherry Files

| | |
|---|---|
| Type | |
| Fruit Qualities | |
| Bloom Time | |
| Cross-Pollinator | |
| Harvest Time | |
| Source | |
| Disease Resistance | |
| Pruning | |
| Soil Amendments | |
| Diseases & Treatments | |
| Pests & Treatment | |

**Notes:**

# Serviceberry Files

| | |
|---|---|
| Type | |
| Fruit Qualities | |
| Bloom Time | |
| Cross-Pollinator | |
| Harvest Time | |
| Source | |
| Disease Resistance | |
| Pruning | |
| Soil Amendments | |
| Diseases & Treatments | |
| Pests & Treatment | |

**Notes:**

# Native Nut Files

| | |
|---|---|
| Type | |
| Nut Qualities | |
| Bloom Time | |
| Cross-Pollinator | |
| Harvest Time | |
| Source | |
| Disease Resistance | |
| Pruning | |
| Soil Amendments | |
| Diseases & Treatments | |
| Pests & Treatment | |

**Notes:**

# Native Nut Files

| | |
|---|---|
| Type | |
| Nut Qualities | |
| Bloom Time | |
| Cross-Pollinator | |
| Harvest Time | |
| Source | |
| Disease Resistance | |
| Pruning | |
| Soil Amendments | |
| Diseases & Treatments | |
| Pests & Treatment | |

**Notes:**

# Native Fruit Files

| | |
|---|---|
| Type | |
| Fruit Qualities | |
| Bloom Time | |
| Cross-Pollinator | |
| Harvest Time | |
| Source | |
| Disease Resistance | |
| Pruning | |
| Soil Amendments | |
| Diseases & Treatments | |
| Pests & Treatment | |

**Notes:**

# Native Fruit Files

| | |
|---|---|
| Type | |
| Fruit Qualities | |
| Bloom Time | |
| Cross-Pollinator | |
| Harvest Time | |
| Source | |
| Disease Resistance | |
| Pruning | |
| Soil Amendments | |
| Diseases & Treatments | |
| Pests & Treatment | |

**Notes:**

 

# Notes

 

# Tropicals

# Avocado Files

| | |
|---|---|
| Type | |
| Fruit Qualities | |
| Bloom Time | |
| Cross-Pollinator | |
| Harvest Time | |
| Source | |
| Disease Resistance | |
| Pruning | |
| Soil Amendments | |
| Diseases & Treatments | |
| Pests & Treatment | |

**Notes:**

# Avocado Files

| | |
|---|---|
| Type | |
| Fruit Qualities | |
| Bloom Time | |
| Cross-Pollinator | |
| Harvest Time | |
| Source | |
| Disease Resistance | |
| Pruning | |
| Soil Amendments | |
| Diseases & Treatments | |
| Pests & Treatment | |

**Notes:**

# Banana Files

________________________________

| | |
|---|---|
| Size | |
| Fruit Qualities | |
| Bloom Time | |
| Cross-Pollinator | |
| Harvest Time | |
| Source | |
| Disease Resistance | |
| Pruning | |
| Soil Amendments | |
| Diseases & Treatments | |
| Pests & Treatment | |

**Notes:**

# Banana Files

| | |
|---|---|
| Size | |
| Fruit Qualities | |
| Bloom Time | |
| Cross-Pollinator | |
| Harvest Time | |
| Source | |
| Disease Resistance | |
| Pruning | |
| Soil Amendments | |
| Diseases & Treatments | |
| Pests & Treatment | |

**Notes:**

# Coconut Files

| | |
|---|---|
| Type | |
| Nut Qualities | |
| Bloom Time | |
| Cross-Pollinator | |
| Harvest Time | |
| Source | |
| Disease Resistance | |
| Pruning | |
| Soil Amendments | |
| Diseases & Treatments | |
| Pests & Treatment | |

**Notes:**

# Date Files

| | |
|---|---|
| Type | |
| Fruit Qualities | |
| Bloom Time | |
| Cross-Pollinator | |
| Harvest Time | |
| Source | |
| Disease Resistance | |
| Pruning | |
| Soil Amendments | |
| Diseases & Treatments | |
| Pests & Treatment | |

**Notes:**

# Fig Files

| | |
|---|---|
| Type | |
| Fruit Qualities | |
| Bloom Time | |
| Pollinator | |
| Harvest Time | |
| Source | |
| Disease Resistance | |
| Pruning | |
| Soil Amendments | |
| Diseases & Treatments | |
| Pests & Treatment | |

**Notes:**

# Fig Files

| | |
|---|---|
| Type | |
| Fruit Qualities | |
| Bloom Time | |
| Pollinator | |
| Harvest Time | |
| Source | |
| Disease Resistance | |
| Pruning | |
| Soil Amendments | |
| Diseases & Treatments | |
| Pests & Treatment | |

**Notes:**

# Guava Files

| | |
|---|---|
| Size | |
| Fruit Qualities | |
| Bloom Time | |
| Cross-Pollinator | |
| Harvest Time | |
| Source | |
| Disease Resistance | |
| Pruning | |
| Soil Amendments | |
| Diseases & Treatments | |
| Pests & Treatment | |

**Notes:**

# Mango Files

| | |
|---|---|
| Size | |
| Fruit Qualities | |
| Bloom Time | |
| Cross-Pollinator | |
| Harvest Time | |
| Source | |
| Disease Resistance | |
| Pruning | |
| Soil Amendments | |
| Diseases & Treatments | |
| Pests & Treatment | |

**Notes:**

# Olive Files

| | |
|---|---|
| Type | |
| Fruit Qualities | |
| Bloom Time | |
| Cross-Pollinator | |
| Harvest Time | |
| Source | |
| Disease Resistance | |
| Pruning | |
| Soil Amendments | |
| Diseases & Treatments | |
| Pests & Treatment | |

**Notes:**

# Olive Files

| | |
|---|---|
| Type | |
| Fruit Qualities | |
| Bloom Time | |
| Cross-Pollinator | |
| Harvest Time | |
| Source | |
| Disease Resistance | |
| Pruning | |
| Soil Amendments | |
| Diseases & Treatments | |
| Pests & Treatment | |

**Notes:**

# Papaya Files

| | |
|---|---|
| Size | |
| Fruit Qualities | |
| Bloom Time | |
| Cross-Pollinator | |
| Harvest Time | |
| Source | |
| Disease Resistance | |
| Pruning | |
| Soil Amendments | |
| Diseases & Treatments | |
| Pests & Treatment | |

**Notes:**

# Pineapple Files

| | |
|---|---|
| Size | |
| Fruit Qualities | |
| Bloom Time | |
| Harvest Time | |
| Source | |
| Disease Resistance | |
| Soil Amendments | |
| Diseases & Treatments | |
| Pests & Treatment | |

**Notes:**

# Pomegranate Files

| | |
|---|---|
| Type | |
| Fruit Qualities | |
| Bloom Time | |
| Cross-Pollinator | |
| Harvest Time | |
| Source | |
| Disease Resistance | |
| Pruning | |
| Soil Amendments | |
| Diseases & Treatments | |
| Pests & Treatment | |

**Notes:**

# Pomegranate Files

| | |
|---|---|
| Type | |
| Fruit Qualities | |
| Bloom Time | |
| Cross-Pollinator | |
| Harvest Time | |
| Source | |
| Disease Resistance | |
| Pruning | |
| Soil Amendments | |
| Diseases & Treatments | |
| Pests & Treatment | |

**Notes:**

# Dragonfruit Files

| | |
|---|---|
| Type | |
| Fruit Qualities | |
| Bloom Time | |
| Cross-Pollinator | |
| Harvest Time | |
| Source | |
| Disease Resistance | |
| Pruning | |
| Soil Amendments | |
| Diseases & Treatments | |
| Pests & Treatment | |

**Notes:**

# Starfruit Files

| | |
|---|---|
| Size | |
| Fruit Qualities | |
| Bloom Time | |
| Cross-Pollinator | |
| Harvest Time | |
| Source | |
| Disease Resistance | |
| Pruning | |
| Soil Amendments | |
| Diseases & Treatments | |
| Pests & Treatment | |

**Notes:**

 

# Notes

 

# Caffeine Plants

# Cocoa Files

| | |
|---|---|
| Type | |
| Fruit Qualities | |
| Bloom Time | |
| Cross-Pollinator | |
| Harvest Time | |
| Source | |
| Disease Resistance | |
| Pruning | |
| Soil Amendments | |
| Diseases & Treatments | |
| Pests & Treatment | |

**Notes:**

# Cola Files

| | |
|---|---|
| Type | |
| Nut Qualities | |
| Bloom Time | |
| Cross-Pollinator | |
| Harvest Time | |
| Source | |
| Disease Resistance | |
| Pruning | |
| Soil Amendments | |
| Diseases & Treatments | |
| Pests & Treatment | |

**Notes:**

# Coffee Files

| | |
|---|---|
| Type | |
| Bean Qualities | |
| Bloom Time | |
| Cross-Pollinator | |
| Harvest Time | |
| Source | |
| Disease Resistance | |
| Pruning | |
| Soil Amendments | |
| Diseases & Treatments | |
| Pests & Treatment | |

**Notes:**

# Coffee Files

| | |
|---|---|
| Type | |
| Bean Qualities | |
| Bloom Time | |
| Cross-Pollinator | |
| Harvest Time | |
| Source | |
| Disease Resistance | |
| Pruning | |
| Soil Amendments | |
| Diseases & Treatments | |
| Pests & Treatment | |

**Notes:**

# Guarana Files

| | |
|---|---|
| Type | |
| Fruit Qualities | |
| Bloom Time | |
| Cross-Pollinator | |
| Harvest Time | |
| Source | |
| Disease Resistance | |
| Pruning | |
| Soil Amendments | |
| Diseases & Treatments | |
| Pests & Treatment | |

**Notes:**

# Tea Files

| | |
|---|---|
| Type | |
| Leaf Qualities | |
| Bloom Time | |
| Cross-Pollinator | |
| Harvest Time | |
| Source | |
| Disease Resistance | |
| Pruning | |
| Soil Amendments | |
| Diseases & Treatments | |
| Pests & Treatment | |

**Notes:**

# Tea Files

| | |
|---|---|
| Type | |
| Leaf Qualities | |
| Bloom Time | |
| Cross-Pollinator | |
| Harvest Time | |
| Source | |
| Disease Resistance | |
| Pruning | |
| Soil Amendments | |
| Diseases & Treatments | |
| Pests & Treatment | |

**Notes:**

# *Tea Files*

| | |
|---|---|
| Type | |
| Leaf Qualities | |
| Bloom Time | |
| Cross-Pollinator | |
| Harvest Time | |
| Source | |
| Disease Resistance | |
| Pruning | |
| Soil Amendments | |
| Diseases & Treatments | |
| Pests & Treatment | |

**Notes:**

# Yaupon Files

| | |
|---|---|
| Type | |
| Leaf Qualities | |
| Bloom Time | |
| Cross-Pollinator | |
| Harvest Time | |
| Source | |
| Disease Resistance | |
| Pruning | |
| Soil Amendments | |
| Diseases & Treatments | |
| Pests & Treatment | |

**Notes:**

# Yerba Mate Files

| | |
|---|---|
| Type | |
| Leaf Qualities | |
| Bloom Time | |
| Cross-Pollinator | |
| Harvest Time | |
| Source | |
| Disease Resistance | |
| Pruning | |
| Soil Amendments | |
| Diseases & Treatments | |
| Pests & Treatment | |

**Notes:**

 

# Notes

 

# Notes

 

# Index

Melanie Holsti is an avid tea lover with a deep appreciation for the simple joys of life.

Whether she's sipping a cup of Scottish Breakfast tea, capturing the perfect shot with her camera, or getting her hands dirty in the garden, she embraces each moment with enthusiasm. As an author and small business owner, she pours her focus and creativity into everything she does.

Her background is as varied as her interests—she's worked in hospitality, retail, livestock farming, writing, and, of course, gardening. Over the years, she has experienced the triumphs and challenges that come with growing food and homesteading in diverse climates, from the harsh prairie of South Dakota to the woods and mountains of Missouri and Tennessee.

In 2018, Melanie and her family took on an exciting project—restoring a charming 1917 Craftsman bungalow. With a lot of hard work and a shared vision, they transformed it into the Ironstone Cottage Tea Room, where guests could gather and enjoy a hot cup of tea or lunch with friends in a cozy, welcoming atmosphere.

These days, Melanie, her husband, and their daughters reside on a small farm in Tennessee, living in a beautiful Federal-style home they built together. Their homestead is shared with an assortment of animals, including a peculiar flock of barn cats, free-ranging chickens, and noisy ducks.

When she's not tending to the garden or caring for her animals, Melanie finds joy in writing books, sampling new and exotic teas, and cheering on her favorite professional hockey team. No matter the season, she's always pursuing her interests and creating a life filled with warmth and, of course, plenty of tea.

www.ironstone tennessee.com

IronstoneTennessee

IronstoneTennessee

amz

amazon.com/author/ironstonetennessee

ironstonetennessee

secret resources web page
Password: IronstoneBooks

Your feedback and reviews are priceless. If you have suggestions for improving my gardening workbook, I'd love to hear them. If you have a moment, please consider leaving a review on Amazon.

*Other Titles by Melanie Holsti*

The Ironstone Garden Planner

The Ironstone Garden Journal

The Fruit, Flower, Herb & Vegetable Garden Planner

The Fruit, Flower, Herb & Vegetable Garden Journal

The Vegetable Files

The Houseplant Files

The Herb Files

The Ornamental Files

Journaling Your Weather

The Ironstone Forager's Notebook

Journaling Your Health (a Collection of Women's Wellness Journals)

Ironstone Prayer Journals

Prayer Team Journals

Tea & Tisane Blending Workbook: A Tea Mixologist's Journal (with Jennifer C. Petersen)

Made in United States
Orlando, FL
06 November 2026

90244042R10129